AF882441

SOTHIS PRESS

Sothis Press aims to completely reimagine the public domain book with an emphasis on universal access to captivatingly designed print editions. We believe that technology will enable every single book ever published to be perpetually accessible in beautiful print form. Sothis is beauty in ubiquity.

www.sothis.press
1 Waterhouse Square, London, ECIN 2ST

A Guide-Book to the Lake Louise Region

Including Paradise Valley of the Ten Peaks, and Neighborhood of Lake O'hara

Walter Dwight Wilcox

SOTHIS PRESS

ISBN 978-1-0368-0421-3

This edition is published by
Sothis Press © 2023. The company
is registered in England and Wales.
The Company Number is 14900389.

A CAMP IN CONSOLATION VALLEY

A GUIDE-BOOK

TO THE

LAKE LOUISE REGION

— INCLUDING —

PARADISE VALLEY, VALLEY OF THE TEN PEAKS, AND NEIGHBORHOOD OF LAKE O'HARA

WITH EIGHTEEN FULL-PAGE ILLUSTRATIONS
FROM PHOTOGRAPHS TAKEN BY THE AUTHOR

WALTER DWIGHT WILCOX, F. R. G. S.
Author of "The Rockies of Canada"

PRESS OF JUDD & DETWEILER, INC.
WASHINGTON, D. C.

PREFATORY NOTE.

The purpose of this little guide-book is to help visitors to the Lake Louise region to spend their time as pleasantly as possible. For the hurried traveler, an outline is given of the most important trips. For those with unlimited time, a considerable number of excursions are described—sufficient to occupy the better part of a month.

The area described embraces the immediate neighborhood of Lake Louise and the four adjoining valleys. This region is one of the most interesting parts of the Canadian Rockies, and is probably not excelled by any equal section in the world for the number and variety of its scenic attractions. In this part of the mountains, hardly ten miles square, are found twenty peaks, fourteen of which are over ten thousand feet and six are more than eleven thousand feet above sea-level. The supreme idea of grandeur in mountain forms, which is not dependent on height alone, is here realized in these precipitous cliffs, which rise from 3,000 to 6,000 feet above narrow, forested valleys. Glaciers two or three miles long, countless snow-fields, rushing torrents and waterfalls, rich forests, and park-like expanses of Alpine flowers compel unbounded admiration in every part of this region, but in one respect especially this part of the Canadian Rockies has no rival, whatever may be the diversity of opinion as to its other charms, and that is in the beauty of its lakes. At least three of these—Lakes Louise, Moraine, and O'Hara—are easily without peers in the Canadian Rockies, while McArthur and Wenkchemna lakes help make up a galaxy of mountain tarns unequalled in any part of the world.

The work is the result of many summers of climbing and scrambling in this the author's favorite part of the mountains. The altitudes are those of the Topographical Survey of Canada, supplemented in a few cases by the author's aneroid readings. The distances are calculations from the author's own rate of walking, and the times required by the various trips are obtained by adding one-fifth to same, so as to give a fairer average for those not in the best of training. No claim is laid to extreme accuracy of detail, and the author invites such suggestions and information as may enhance the value of future editions. The illustrations have been selected with the idea of giving the best general idea, as well as to serve as a souvenir of the more interesting views.

W. D. W.

WASHINGTON, D. C., *July*, 1909.

PART I.

LAKE LOUISE AND ITS IMMEDIATE VICINITY.

The following trips are recommended for those having only a limited time at their disposal:

First day (morning).—Ascent to upper lakes, and, if possible, to top of the Little Beehive (page 12). Afternoon.—Walk around Lake Louise to end, or as far up valley as the corduroy bridge (page 9).

Second day.—Ascent to Saddleback, with view of Mount Temple and Paradise Valley (page 17). Afternoon.—Rowboat around lake.

Third day.—To Moraine Lake, in the valley of the Ten Peaks, and, if possible, side excursion to Consolation Valley (page 29).

Fourth day.—Paradise Valley to end, returning by way of Giant Steps Falls (page 21).

For those spending considerable time the following trips are suggested: 1. To Lake O'Hara and Lake McArthur from Hector. 2. Over the Mitre Col and return by Paradise Valley. 3. To Lakes O'Hara and Hector or Field by way of Abbot Pass. 4. Ascent of Mount Victoria.

Laggan, the railroad station for Lake Louise, is a small settlement in the Bow Valley 5,037 feet above sea-level, 36 miles northwest of Banff. It is a divisional point of the Canadian Pacific Railway, and here also the time changes from Mountain to Pacific, or *vice versa.* Near the station is a granite shaft erected to the memory of Sir James Hector, who, in connection with the Palliser Expedition, discovered, in 1858, the Kicking Horse Pass, by which the railroad crosses the highest range of the Rockies. The surrounding forests were destroyed by fire many years ago, but a vigorous growth of jack-pine is now rapidly covering the bare hillsides. Vehicles and ponies meet all trains to convey passengers to the chalet at Lake Louise. The distance by new road is about three miles, but the time required for the trip is nearly one hour, owing to the constant ascent, which amounts to a total of 633 feet.

The *Bow River,* a swiftly rushing glacial stream, is crossed by an iron bridge. Near this is a small spring of excellent water. The carriage road follows the Bow River a short distance, and then makes a gentle ascent through pine forests, which soon open up and disclose the torrential outlet stream from Lake Louise, on the left. Following this, on the west bank, in constant and rather steep ascent, the road crosses the stream (one mile) by a log bridge, and a long detour is made through the woods in order to avoid the excessive grades of the old road. Those who prefer to walk, however, should follow the old road along banks of the stream to the chalet, which may be reached on foot in 45 minutes from the station.

Excellent views of the mountains, especially of Mount Temple, are to be had from the new road as it winds through light forests. A sharp turn is made (1½ miles), and at two miles the stream is again crossed and followed the remainder of the distance to the chalet. Not far from the chalet the electric-lighting plant is passed, and a few hundred yards above this there is a dam where the water is collected for power. The grade soon diminishes, and the noisy stream is hushed as it flows in a broad channel through heavy forests. Glimpses of snow-fields are caught amongst the tall and slender spruce trees, and the blue-green waters of Lake Louise are seen as the chalet is reached.

**Lake Louise Chalet,* delightfully situated at the northeast end of Lake Louise, has accommodations for 175 to 200 guests. There are 120 rooms, of which 83 have baths connecting. Though this hotel is located in the heart of the mountains and the surrounding region was an unexplored and almost unknown wilderness as late as 1890, the visitor may now enjoy such modern comforts as electric lights, steam heat, running water, and an excellent cuisine, amid homelike and restful surroundings. The original log chalet, built in 1890, burned down in 1892, and a small frame structure, capable of accommodating a dozen guests, was built in 1893. Additions were made from time to time to provide for the rapidly increasing number of visitors. In the summer of 1893 there were about 50 visitors to Lake Louise, and in 1908 there were more than 5,700. The original frame chalet is now incorporated in the general sitting-room and music-room. Large, open fireplaces, where pine and spruce logs burn brightly, temper the chill of evening and early morning and give a cheerful aspect to the public rooms. The verandas and large windows on the lake side of the building give unrivaled views.

Rooms, American plan, $3.50 to $5.00; with bath, $4.50 and upward. Conveyance between Laggan and chalet, 50 cents each way; trunks, 75 cents for round trip; hand baggage free. Guides and ponies for excursions in the vicinity or for short camping trips and Swiss guides for mountain climbing may be arranged for at hotel office. Sleeping-car reservations, telegrams, and telephonic messages to Field and Banff, at office.

** *Lake Louise,* one of the most beautiful lakes in the world, hardly known before 1890, when the first chalet was built, is now justly celebrated for an unusual combination of Alpine grandeur with beauty of forested slopes and richly-colored water. The lake is 5,670 feet above sea-level, 1¼ miles in length by one-fourth mile in width, and opposite the rock slide is 230 feet deep. The lake was visited as early as 1882 by Mr. T. E. Wilson, and again in 1886 by Dr. Edwards, a British scientist, who named it after his eldest daughter. The earliest available description of this spot is given in "Among the Selkirk Glaciers," by Dr. Green, who visited the lake in 1888.

Mount Victoria (11,355 feet), the most conspicuous feature of the view, is a lofty wall of rock running at right angles to the line of vision, closing in the valley and forming a part of the continental divide between the waters that flow into the Atlantic and those that drain into the Pacific. Its base is a precipice about 1,500 feet high, while its upper half is almost completely covered with snow and glaciers from 200 to 250 feet thick, where they are crowded over the upper part of the precipice. In hot weather the thunder of avalanches is frequently heard, as immense masses of ice break off and fall to the glacier below. As the sound requires about 20 seconds to reach the chalet, these avalanches are usually ended before the attention is called to them. To the left, and somewhat nearer, a part of the helmet-shaped Mount Lefroy is visible.

On the right may be seen, in relief against the sky, the two rocky and nearly bare summits of Mount Whyte and Mount Niblock. Still farther to the right is the cone-shaped Mount Saint Piran, and to the left of this, and much lower, is the dome of rock called the Beehive. To the left of the lake the imposing precipices of Mount Fairview rise in an impressive wall of rock, at the base of which rests a vast slide of rock partially filling the lake. It represents the erosive action of past ages on the upper cliffs of the mountain. Visitors should be cautious in approaching this and similar slopes on account of the frequent rock-falls.

It is very difficult, at first sight, even for those experienced among mountains, to appreciate the true scale and perspective of such a scene. To get a better idea, the visitor should gradually accustom himself to the following distances: It is one mile and a quarter to the end of the lake, and about two miles in a straight line to the nearest moraine debris, where a stream is seen beyond the water. Above this the glacier extends two miles farther to the lowermost cliffs of Mount Victoria, to reach which from the chalet would require about two hours' hard walking. The summit of Mount Victoria is about five miles distant in a straight line, and rises more than a mile in vertical height above the lake.

The nearly vertical lower cliffs of Mount Fairview on the left rises between 1,700 and 1,800 feet above the lake. If the lake were deep enough to submerge three such cliffs, piled one on top of the other, the bottom would be several

LAKE LOUISE AND MOUNT VICTORIA

hundred feet above sea-level. The top of this cliff is about on a level with the ice of the Victoria glacier as seen at the base of Mount Victoria. Another such cliff piled on top of the first would reach the top of the hanging glacier, and still a third cliff piled on top of the other two and placed in front of Mount Victoria would leave fully 600 feet of the latter's snowy crest towering above. The Beehive, on the right, is about the same height as this cliff, while Mounts Whyte and Niblock, though rising over 4,000 feet above the lake, are one-third of a mile lower than Mount Victoria. In rainy weather the level of the clouds makes these facts strikingly apparent.

***Excursion to end of lake* (by boat, 20 minutes; on foot, 25-30 minutes) is highly recommended, both by boat and on foot, as one of the most satisfactory of the easier trips at Lake Louise. The trail commences on your right, or north shore of the lake, near the boat-house, and follows the water closely throughout. The view changes constantly and is highly interesting. There is a rustic pavilion (about ¼ mile) commanding a fine view, and shortly beyond this a small rivulet enters the lake, which bursts out of the ground in a beautiful spring about 80 feet above. The snowy dome of Mount Lefroy, the imposing peak of Mount Aberdeen, and the nearer spires and minarets of a wall-like spur from Mount Fairview appear as you progress. Near the lake end there is a very good view of Mount Lefroy, seen through the constricted and canyon-like valley, but even this imposing mountain cannot arouse greater admiration than the frowning cliffs that now appear directly ahead. Excessively steep, and even perpendicular in places, for several hundred feet of vertical height, these cliffs show an unusual variety of color, partly due to the several kinds of stone of which they are composed and to the iron and other minerals with which the dripping water has stained the rock walls in vertical bands. Crossing an open slide, decorated in midsummer by false forget-me-nots and other Alpine plants, the trail leads up to the base of the cliffs. Here, in the partial gloom of the forest, the cliffs above can be more thoroughly studied. The narrow shelves of rock, three or four hundred feet above, hardly seem adequate support for the clinging spruces which are now seen directly overhead. Descending through a massive pile of rocks, which have fallen in past ages from the cliffs, the trail reaches a flat meadow at the lake end (1½ miles), filled with cotton grass and sedges. This nearly level meadow, extending for nearly a quarter mile up the valley, represents the former lake bed, now filled in with mud and gravel from the glacier. In hot weather the discolored stream may be seen extending a long way into the lake.

The Victoria Glacier and cliffs of Mount Victoria.—From end of lake the trip may be extended (20 minutes) over a good trail to a footbridge, near end of glacier, or (1½ hours) over more difficult trail to edge of the middle glacier and cliffs of Mount Victoria. The trail, crossing the ramifying branches of the inlet stream, now becomes faint until in a few hundred yards it enters a deep forest on the north side of a brawling torrent and ascends rapidly in steep pitches, crossing open slides filled with cow-parsnip and false hellebore, to reach the lowermost of the ancient bear-den moraines, one of which is crossed by a rustic footbridge (2⅓ miles from chalet). After crossing this bridge the trail winds amongst the rocks and continues across a gravel wash to a point near end of the glacier, where the torrential stream is hemmed in by a low, rocky cliff. To avoid this the trail turns to the right and seeks an outlet onto the well-defined moraine of the glacier. From this point on the trail becomes fainter, but there is no difficulty in following a fairly good route over these moraine ridges for about three-fourths of a mile farther, when the walking becomes much easier and pleasanter to (5 miles) base of Mount Victoria.

The Victoria Glacier, three miles long by one-half mile wide, can be explored from this part of the valley. A Swiss guide should be employed to point out the interesting sights, such as glacial tables, where the great boulders are seen supported on pillars of ice, glacial streams, the curious surface lakelets, moulins, crevasses, etc. The trip over the ice should be extended to the base of the cliffs of Mount Lefroy (well worth seeing at close range) and the base of Abbot Pass, formerly known as the Death Trap.

LAKE LOUISE AND MOUNT LEFROY

The Lefroy Glacier is a branch of the Victoria Glacier, lying between Mounts Lefroy and Aberdeen, and is over a mile long by nearly half a mile wide. It is hemmed in by the Mitre Mountain at its upper end, and by the precipices of Mount Lefroy, which here rise in an unbroken wall about 2,500 feet above the ice. If the weather is warm the visitor may be rewarded by seeing one of the inspiring avalanches that descend either from the hanging glaciers of this mountain or from those of Mount Victoria.

Dr. Scherzer gives the following information relative to these glaciers: The Victoria Glacier originates at Abbot's Pass, flows due north for a mile between the precipitous walls of Mounts Victoria and Lefroy, makes an abrupt turn to the northeast, and pursues a straight course for another two miles before wasting away in the Lake Louise Valley. The lower 1½ miles present a remarkably even surface, so that it may be ascended by the most inexperienced. The thickness of the ice opposite the Lefroy tributary is probably from 500 to 600 feet. The calculated annual flow of ice down the valley ranges from a few inches near the edges of the glacier up to about 66 feet near the center of the stream. According to measurements carried on by the Messrs. Vaux, and later by Dr. Scherzer, the glacier retreated an average of 14½ feet yearly from 1899 to 1905.

**The Mitre Col* (8,500 feet), lying between Mount Aberdeen and the Mitre, may be ascended (guides advisable) by way of the Lefroy Glacier and descent made into Paradise Valley, passing the Giant Steps Falls and returning by trail to chalet. This all-day trip, with its glacier and snow work, gives a great variety of climbing and affords a constant succession of inspiring and beautiful views. The Mitre is ascended by way of this pass, but is a cliff-climbing problem of considerable difficulty only hitherto solved by a very few parties.

***The ascent of Mount Victoria* (11,355 feet), recommended as the most interesting of the accessible Alpine peaks near Lake Louise, is made by way of Abbot Pass. This is a high col or snow pass (9,588 feet) between Mount Victoria and Mount Lefroy, named for Philip S. Abbot, who lost his life here in 1896 while attempting the ascent of Mount Lefroy. (See Appalachia, Vol. VIII, No. 2.) The rope is usually put on shortly after rounding the rock buttress of Mount Lefroy, from which point the snow-covered glacier slopes lead up steeply into a narrow gorge hemmed in by the precipices and hanging glaciers of Mount Victoria. Danger of avalanches in this particular part suggested the original name of the "Death Trap," still applied locally to the lower part of Abbot Pass. From the summit of this pass a remarkable view is unfolded on both sides, but more particularly towards the south, where Mount Biddle and other lofty and sharp peaks tower above a sea of snow and rocks, making an awesome but inspiring panorama. The descent to Lake O'Hara by way of Lake Oesa is steep in upper part, but soon becomes less difficult. The ascent of Mount Victoria is interesting and varied from the summit of Abbot Pass. A rather steep rock buttress is climbed just above the pass leading out to a rocky arête. This is now followed either on the rocks or the crest of a snow ridge to the summit, which, however, lying a mile or more in a straight line to the north, is only reached after several hours' work. The difficulty of this ascent along this high, exposed ridge depends, of course, largely on weather conditions, and the Swiss guides will be able to advise fully in regard to same. The variety of work required and the constantly interesting panorama make Mount Victoria one of the most advisable climbs, especially as the entire ascent may be made with a comfortable hotel as a base. The north arête of Mount Victoria is defended by gendarmes and vertical precipices, and the ascent has not been made as yet from this side.

**Mirror Lake and Lake Agnes.*—(To Mirror Lake, 30 minutes; to Lake Agnes, 45 minutes. Horses may be taken nearly to Lake Agnes, the higher of the two lakes, and last part made on foot.) This trip, involving a climb of 1,000 feet to Mirror Lake and 1,200 feet to Lake Agnes, is very interesting and gives a good general idea of the surrounding mountains.

The trail commences north of the chalet, a little to the right of the trail round the lake, and ascends in sharp grade, almost without interruption for first half mile, through a light forest of spruce and pine. The shrubby bushes are mostly mountain rhododendron, flowering throughout July and August, the bushes at higher altitudes being correspondingly later. The trail passes a reservoir for the chalet water supply, and at about 600 feet elevation turns to the left in a nearly level traverse, and is thereafter of much easier grade. The hardest and least interesting part of the climb is over at this point. Gradually ascending, the trail enters a heavier forest, and winding amongst some large moss-covered stones, divides, the right-hand trail ascending sharply to come down on Lake Agnes from above, the left-hand branch continuing in easier grade to Mirror Lake, and thence to Lake Agnes from below. Either route may be taken, but the left-hand trail is advised, as it is easier and the scenery is revealed in a more interesting manner. In a short distance from the division of the trail *Mirror Lake,* a round pool 100 yards across, is seen on the right, fed by the waters of Lake Agnes, and seeking an underground outlet to Lake Louise. From its secluded position in the forest and its small size, the water is rarely disturbed by the breezes, and hence its surface usually reflects the surrounding cliffs and trees in a mirror-like surface. A rounded dome of rock called the *Beehive,* from its peculiar shape, is the most striking feature from the shores of Mirror Lake. The summit of this is about 750 feet above the lake and its upper cliffs are nearly perpendicular.

The trail continues to the left of the lake and ascends sharply for a short distance to an opening amongst the trees, whence a fine view may be obtained of the snowy crest of Mount Aberdeen to the south. The Lyall's larch, a tree resembling the tamarack, is now seen on all sides, and is an indication of the increasing altitude. The eastern cliffs of the Beehive and the broken rocks which have fallen from them in past ages are imposing from this point. As the trail turns to the right it begins to skirt the base of the Beehive, showing Mirror Lake below and giving a very extensive view across the Bow Valley and the mountains of the Sawback Range to the east. The view rapidly increases in interest as the trail ascends in sharp zigzags, allowing a glimpse of the outlet stream from Lake Agnes, which falls in a pretty cascade over rock ledges to the right, and finally reaches a steep flight of steps, ascending which the visitor stands upon the heath-lined shore of—

Lake Agnes.—This lake, about one-fourth of a mile long, is 6,875 feet above sea-level, or almost exactly 1,200 feet above Lake Louise. It was named for Agnes Knox, who visited it September 15, 1890, the first woman to reach the lake, which must have been a very trying ordeal through the trailless forests. The lake occupies the foot of a rather barren, cirque-like valley between the Beehive and Mount Saint Piran, and overhung by the cliffs of Mount Whyte and Mount Niblock. The north exposure and great altitude of this little lake cause the snows of winter to linger round these barren shores till midsummer, and the winter coat of ice sometimes does not melt away till the first week in July. Near the outlet stream is a rustic cabin erected as a shelter against sudden storms.

From near the cabin the trail turns to the north, and, ascending slightly, crosses the face of a steep, rocky slope, whence, amongst the clinging trees, very inspiring views may be had of the cliffs of the Beehive and distant mountains. Traversing a bare slope, whence the snowy crest of Mount Lefroy appears above the Beehive, the trail now descends in sharp zigzags amongst open larch groves to join the trail near Mirror Lake.

Many variations and extensions of this excursion may be made by those desirous of getting a further insight into the truly remarkable scenery to be enjoyed from near-by points, of which the following, to the summit of the Little Beehive, is strongly recommended.

From the highest part of the trail, as it crosses the bare slope mentioned a few lines above, may be seen a line of cliffs to the north, about one-fourth of a mile distant and a few hundred feet higher, towards which an easy route may be selected over the somewhat rocky and heath-covered slope, gradually ascending so as to come out above the cliffs. A grove of larches is entered and a slight descent made, only

MOUNTS ABERDEEN AND LEFROY FROM THE LITTLE BEEHIVE

to rise again in another hundred yards. Here it is advisable to keep near the edge of the cliffs, so as to get a better view of the scenery, which increases in interest till the rolling, larch-crowned summit of the Little Beehive is reached. From the edge of this cliff there is a—

** *Magnificent view* of the snowy summits of Mounts Victoria, Lefroy, and Aberdeen, and even distant Hungabee, in Paradise Valley, now looming up in inspiring grandeur across the deep, forested valley of Lake Louise, the banded cliffs of the Beehive and the sparkling Mirror Lake at its base, while farther to the left is the giant dome of Mount Temple and Mount Fairview. To the east and north lies the open sweep of the Bow Valley, showing the mountains from near Banff around to the north, where may be seen (by walking a hundred yards to the other side of the ridge top) the Pipestone Valley, Mounts Molar and Hector, the upper Bow Valley and Hector Lake, with the distant and snowy peaks near the source of the Bow River. On a bright day it is advisable to take luncheon in this spot and spend several hours enjoying this almost unrivalled panorama. To those who have been stimulated by this scene to make further efforts the ascent is recommended of—

Mount Saint Piran (8,681 feet, named for town in Cornwall, birthplace of W. J. Astley, the first manager of the Lake Louise chalet). The mountain seems deceptively near, but as a matter of fact the top of the Little Beehive is only a little more than half the height from Lake Louise to the summit of Saint Piran. There is no difficulty, however, for vigorous walkers of either sex, provided with nailed shoes, in reaching the rocky, nearly barren summit (40 to 50 minutes) from the Little Beehive. Route is obvious from below.

* *The view from summit* is magnificent, being an extension of that from the Little Beehive, and includes, besides, a glimpse of the deep gorge or short valley north of Mount Niblock, with a small lake far below. The altitude here is almost exactly 3,000 feet above Lake Louise, and slightly more than half the height of Mount Victoria above same. This level also very nearly marks the extreme upper limit of vegetation. A few species of hardy Alpine plants, mostly

composites, are visited on bright days by rare kinds of butterflies and insects, only found elsewhere in the Arctic regions. The descent may be varied by climbers of a little experience by following the south slopes in a direction towards Mount Niblock and coming down the Alpine Valley, which ends at Lake Agnes.

The Beehive (7,430 feet) may be ascended by skirting Lake Agnes on the left or south side, using caution amongst the unbalanced and somewhat treacherous rocks along the shore, until a grassy and wooded slope is reached near the end of the lake, when the very steep but safe (shoes should be hob-nail'ed) ascent is made, 555 feet, to the pleasant, larch-covered and nearly flat summit. From this point—

* *The view* of Mount Lefroy and the upper part of the Lake Louise Valley is magnificent and awe-inspiring. The visitor should cautiously approach the nearly vertical cliffs on the east and north sides to obtain an idea of their height and steepness. From the flat top of the Beehive the trip may be extended in a very interesting climb along the lower ledges of Mount Whyte, which may be traversed in a direction towards Mount Victoria on a nearly level route, much facilitated by trails made by the wild goats. The views are interesting, and divide the attention with the cliffs on the right and the beautiful Alpine flowers which grow amongst these high ledges. The descent may be made directly to the Lake Louise Valley, but this is a rough scramble at best. The route should be studied out in order to avoid, as far as possible, the scrubby brush and ledges which are the only difficulties.

The Devil's Thumb (8,065 feet), a rocky cliff above the Beehive, may be ascended easily from the above described ledges or directly from the lower valley, but the route should be chosen in advance, as it is possible to lose the bearings when actually on the mountain side.

* *The view* is similar to that from the Beehive, but the awe-inspiring cliffs on the Lake Agnes side are the most interesting feature.

Ascent of pass between Mount Saint Piran and Mount Niblock via Lake Agnes, and descent of the next valley to

Temple

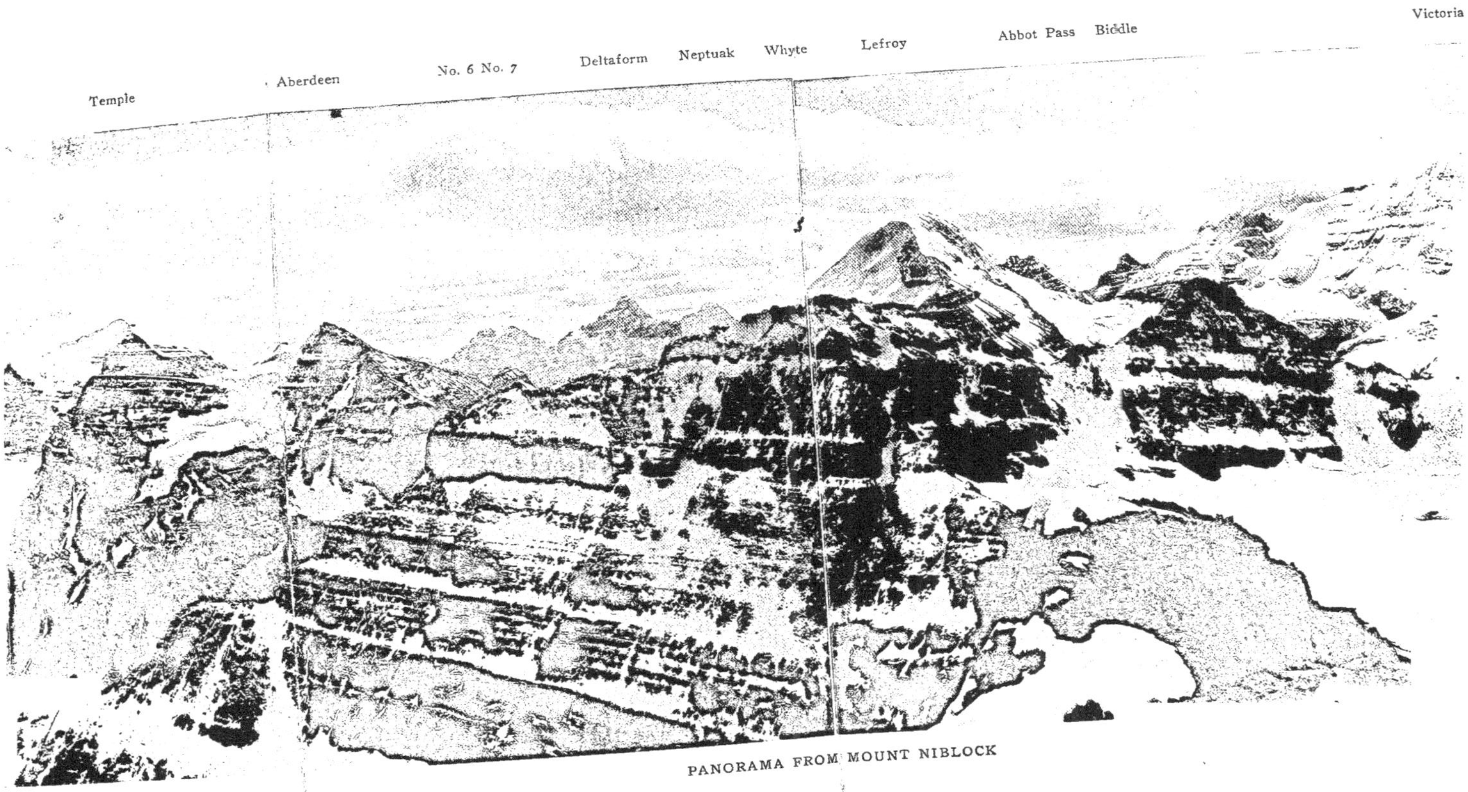

PANORAMA FROM MOUNT NIBLOCK

the north, and return by Hector trail to the chalet. This rather rough trip, requiring 4 to 5 hours, is recommended only to those spending a long time at Lake Louise. The ascent to Lake Agnes is made in the usual way, and the lake shore is skirted over the broken stones and boulders on north side of lake. The general trend of the valley is followed with ease, turning gradually to the right, and an ascent made towards the lowest part of the depression between Mounts Niblock and Saint Piran. The snow patch on north side of the pass, somewhat steep in upper part, must be negotiated with care when frozen hard, after which there is no difficulty in the descent of the valley to a small lake (5,900 feet), 150 yards across, overhung by the gloomy precipices of Mount Niblock to the north, which latter rise nearly 4,000 feet above this rarely visited and lonely tarn. The view is impressive but forbidding. The descent of the outlet stream, which soon becomes a succession of cascades and pretty falls through heavy timber, is trying and rough. The trail is found at length at a point about two miles from the chalet. The great number of intersecting trails and old roads of the last mile are confusing, but by taking every turn to the right the chalet may be reached without loss of time. If this excursion is reversed, the services of some one familiar with the trail should be secured to accompany the party for the first mile or so, and the location of the upper valley containing the lake should be clearly in mind in order to avoid mistakes and possible failure.

The Saddleback, an Alpine meadow about 1,800 feet above Lake Louise, lying south of Fairview Mountain, from which there may be obtained a magnificent view of Mount Temple and Paradise Valley, may be reached in an hour from the chalet, and is probably the finest of all the near-by excursions.

The trail starts from the east end of the chalet, and, crossing the outlet stream by a corduroy bridge, whence a fine view of the lake is obtained, plunges at once into a deep and mossy forest, and begins a rapid ascent of the lower slopes of Fairview Mountain. The trail divides (6 minutes), the one on the left being the main trail to Paradise Valley, the Valley of the Ten Peaks, and Moraine Lake. The trail to the right is followed, which, ascending rapidly, eventually turns to the left and crosses the open track of a former snow-slide (18 minutes; about one-third of the distance to the Saddleback), whence a first view is obtained of the open Bow Valley and the distant mountains of the Pipestone and upper Bow Valleys. A much nearer view is also obtained of Fairview Mountain, the cliffs of which now loom above on the southwest. The trail ascends in very sharp pitches and frequent zigzags through spruce and larch groves. Upon reaching an open flank of the mountain (40 minutes) there is a view of the top of the Saddleback lying to the south, appearing deceptively near, this point being only two-thirds of the way from the chalet to the Saddleback. The enlarging view and beautiful Alpine flowers are a partial compensation for the somewhat arduous final part of the climb. The trail, after a long traverse, makes the final ascent in several sharp zigzags through a grove of fine larches. At the second turning the snowy crest of Mount Temple appears (53 minutes), rising above the green slopes of the meadow, brilliantly white and inspiringly beautiful in clear weather. As the slope falls away the trail becomes indistinct in the grass and heaths which here make an Alpine meadow, typical of the higher Canadian Rockies. The visitor is very likely to see one or more marmots or siffleurs and hear their piercing whistles while approaching this beautiful spot, and with care and patience in arousing their curiosity they may be approached quite closely and photographed. There is a rare chance also of seeing a herd of Rocky Mountain goats, either on the south flanks of Fairview Mountain or near the base of Mount Sheol, as you look into Paradise Valley. Somewhat above the lowest part of the pass there is a log cabin (60 minutes), placed here as a shelter in time of storm. Following a faintly marked trail from near the cabin in the direction of Mount Temple, and descending about 50 feet through the larch trees, a small cliff is reached where there may be obtained an

***Excellent view of Mount Temple and Paradise Valley.* This inspiring view, of its kind probably not excelled on this continent, comprises the massive glacier-crowned Mount

MOUNT TEMPLE FROM THE SADDLEBACK

Temple (11,626 feet), one of the highest peaks of the Canadian Rockies, rising 6,000 feet in almost vertical precipices out of the green, forest-clad Paradise Valley. At the base of the mountain is *Lake Annette,* a sparkling pool, clear and green as an emerald. The milky, glacial torrent, descending from the ice fields at the valley end, appears like a silver thread winding amongst the forest trees 2,000 feet below. The dark and frowning precipices of Mount Sheol hem in the valley and shut off the view on the right.

Saddle Mountain, a rocky summit lying to the east, may be reached in 20 minutes from this point without difficulty, and is recommended as giving a good view of Pinnacle Mountain, the Bow Valley towards Banff, and the remarkable cliffs of the mountain itself. The upper or highest part of the ridge is selected as a route, and the only part requiring caution is amongst the large, loose stones near the summit. The climber should endeavor to get a view down the cliffs on the east side of this mountain, approaching the verge of the vertical precipices with care.

Fairview Mountain, which from the Saddleback resembles a conical pile of debris, may be ascended in one hour from this point. A trail commences near the cabin and winds up the southeast face of the mountain, but the visitor may select almost any route on this side of the mountain over as far as the skyline on the right. This climb, though somewhat monotonous, is entirely free from any difficulties.

**The view from the summit* (9,001 feet) is very good, especially of Mounts Victoria, Aberdeen, and Temple. The panorama on the other sides comprises an immense sweep of the Bow and Pipestone valleys.

A somewhat longer, though very interesting, walk may be had by returning to chalet by way of Paradise Valley. The descent is made from the viewpoint near the cabin (a rather rough scramble) to reach the trail in Paradise Valley (30 to 40 minutes), near the base of Saddle Mountain, from which point to the chalet will require about 1½ hours.

PART II.

PARADISE VALLEY.

Paradise Valley, so named from its unusual beauty, is the first valley to the east of Lake Louise. It is about six miles long by a mile or more in width, and, running northeast, opens into the Bow Valley. About 5,900 feet in altitude at its opening, the level of the valley floor gradually ascends to over 7,000 feet at moraine of the Horseshoe Glacier. Hemmed in by precipitous mountains, three of which are over 11,000 feet high, namely, Mounts Temple, Hungabee, and Lefroy, adorned with rich, primeval forests, waterfalls, glaciers, and other natural attractions, this easily accessible valley is well worthy of a visit. The upper part, owing to its altitude and north exposure, is usually covered with snow till the middle of July, and, if possible, a visit later in the season is advisable.

The lower end of the valley is reached by trail (on foot or horseback) in less than one hour from the chalet. The trail (excellent for the first 3 miles) crosses bridge at outlet of Lake Louise, and, ascending through deep forests, divides (6 minutes—175 feet above Lake Louise), the right-hand trail ascending to the Saddleback. The left branch, gradually ascending, reaches (25 minutes) the path of a snow-slide from the Saddleback, which it follows in a very steep ascent till it attains a height of 375 feet above Lake Louise. From here on trail gradually descends and (55 minutes—about 3 miles) reaches point where Paradise Valley trail branches off to the right or south. The approach to junction of this trail may be recognized by a sharp turn where trail crosses a shallow, ravine-like depression 200 yards before Paradise Valley trail is reached. About one-fourth mile further on, the main trail, continuing east, crosses the Paradise Valley stream by a corduroy bridge to reach the valley of the Ten Peaks.

An alternative route thus far is by carriage road to the Paradise Valley stream. Near the bridge over this a steep trail commences and ascends along west bank of stream one-fourth mile to upper trail. From here follow main trail back one-fourth mile to junction with Paradise Valley trail.

The Paradise Valley trail, from point of junction with main trail, leads southward through heavy forests, and (4 minutes) allows the first view of the triangular peak of Mount Hungabee at head of the valley. At length the nearly level trail, skirting the swamps at the valley entrance, becomes rough (4 miles from chalet) as it crosses rock slides from Saddle Mountain on the right. The scenery now approaches a *climax of grandeur* when, after traversing some virgin forests of spruce and crossing a small, mossy stream (5 miles) that descends from the valley between Mount Sheol and Fairview Mountain, the sound of the now rapid creek is heard and openings in the forest permit views of the surrounding mountains. To the north is Saddle Mountain and the conical pile of Fairview Mountain. Above long slopes of grass, the imposing cliffs of Mount Sheol rise on the right, richly colored red and yellow, and beautified by narrow ledges with clinging trees. To the south is Mount Temple, crowned with a glacier of matchless purity and whiteness. The trail, following the Paradise stream closely for some distance, affords excellent views of the cliffs of Mount Temple, which here rise more than a mile in vertical height above the valley. The actual summit is nearly 6,000 feet above the trail, and the wall of ice above the cliff is from 200 to 300 feet thick. Crossing many openings in the forest made by snow-slides, the visitor arrives (2 hours 20 minutes) at the 1907 camp of the Alpine Club of Canada. The trail crosses stream (bridge), and after half a mile recrosses (footbridge).

The two streams making the Paradise Valley creek here unite, one coming from meadows in the extreme north-western side of the valley and descending in a fine cascade called the *Giant Steps Falls* (one-half mile above), and the other from similar meadows on the other side of the valley. Trail now rapidly ascends a ridge between the two streams, through spruce forests, allowing a fine view of rapids on the left until, after a sharp ascent, open meadows and larch groves diversify the landscape. The scenery here approaches a *second climax of beauty,* and throughout the next mile there is a constant succession of excellent views in every direction. The trail becomes faint in places as it crosses the open meadows on west side of stream, some distance above it, and finally ends near the glacial moraine in the very center of the valley.

The open, upper part of Paradise Valley, during the months of July and August, is a veritable *Alpine flower garden* of great beauty. The rolling ridges, crowned with larches, are masses of color—red, yellow, white, and purple—when the painted cups, arnicas, and asters are at the height of their glory. The two swampy meadows on either side of the central ridge which divides the upper valley are filled with equisetums and masses of white cotton grass. On every side are mountains of varied form, making a great encircling amphitheater or *cul de sac,* with a narrow opening down the valley as the only apparent outlet. Taking up the peaks in detail, the triangular rock pyramid at end of valley is Mount Hungabee (11,447 feet). On the right is a lesser peak and the long wall of Mount Lefroy forming an imposing precipice. Separated by a glacier is the difficult Mitre Mountain, showing two peaks from this point of view. On the other side of the Mitre Pass, to the north, is a high spur of Mount Aberdeen, beyond which the lower slopes of Mount Aberdeen and Mount Sheol are seen extending down the valley. Starting once more with Mount Hungabee, there are seen on the left three lesser peaks, the first of which is a part of the same ridge. Between the rounded dome to the south and the easily recognized Eiffel Peak, with its cleft, tooth-like summit, is the Wastach Pass. To the left of Eiffel Peak is the remarkable Pinnacle Mountain, which has baffled all attempts to climb it, separated by the Sentinel Pass from Mount Temple, now showing a remarkable resemblance to Mount Stephen at Field.

Excursions in upper part of Paradise Valley.—There are three excursions to be made from this part of the valley, one at least being recommended as an interesting enlargement of the usual visit. Taking them up in order of their advisability, they are:

1. *To Giant Steps Falls,* a very beautiful trip, walking only rough in places, and requiring only half an hour more than to return by the trail as the latter is taken up at union of the two streams, near footbridge.

2. *To summit of Wastach Pass,* a mountain pass of average roughness, giving a fine view into upper part of Valley of the Ten Peaks, the climb requiring one hour to summit and considerably less to return. This trip is not for the totally inexperienced climber.

3. *To lower part of Sentinel Pass, with view of Pinnacles;* a rough trip through woods and rock slides; very interesting for a strong walker, requiring from 1 to 2 hours, according to extent of trip taken.

***Giant Steps Falls,* a beautiful cascade or succession of falls, remarkable for the peculiar formation of the rocks, though of no great height or grandeur, is in many respects unsurpassed for its size. They may be reached as follows: From the trail end near the moraine (if horses are taken, they should be sent down—1½ miles—to footbridge at junction of the two streams), near some table-like rocks covered with heaths and trees, or from any part of the upper valley, proceed northwestwards, crossing the low ridge and descending to the flats at base of the Mitre. Some large trees, apparently broken off by a strong wind made by a snow-slide from Mount Lefroy, are first passed. The meadows are crossed, or rather skirted on the right, and the stream followed as it changes from a sluggish, meandering brook, hemmed in by clay banks, to a rapid torrent, flowing over a bed of solid rock. On the right there is a most attractive open grove, with banks of heaths edging the almost level

GIANT STEPS FALLS

stone pavement. The water glides for one-eighth mile in sparkling ripples over the stone floor, ever falling against the left bank, and then forced back again by projecting points of rock. The peculiarly hard and homogeneous rock is polished by former glacial action, the striations and grooves of which are clearly apparent on every side. In the afternoon sun, on a bright day, the gliding waters seem like a stream of molten silver. Suddenly the torrent commences a succession of leaps and falls over immense rectangular blocks of stone, and a succession of truly remarkable views are to be obtained from the banks as the water descends these giant steps of solid rock. The falls are at their best on hot days and in afternoons when the glaciers are melting most rapidly. A swampy meadow, swept clean of trees by a snow-slide, is crossed and the stream followed (one-half mile) to junction of streams, where the trail is found.

**Wastach Pass* (from Indian word meaning "beautiful"—8,338 feet) may be reached in one hour from upper part of valley. Lying south of Eiffel Peak, this pass gives an excellent view into the upper Valley of the Ten Peaks, and is a popular route to same. Cross the upper meadows by keeping near or on the forest-covered slopes of the moraine of the Horseshoe Glacier. Follow the last large stream on the south side of valley a short distance, then cross same and ascend any of the numerous gulches towards the pass. The country is open and attractive, beautified by flower-lined vales, and many small streams descend the horizontal ledges, making the cliffs drip and glisten in the sunlight. As the lower part of the pass is approached the vegetation ceases, and the ascent is made over broken stones and rock-slides. The last part of ascent, somewhat steep, should be made on left side of the pass, where goat trails facilitate the climbing. **The view* into the Valley of the Ten Peaks, including Mount Deltaform and four other peaks of the Wenkchemna Range, is one of the wildest and most desolate in the mountains.

* *The Pinnacles* or rock towers on north side of Pinnacle Mountain may be well seen from the lower part of the Sentinel Pass, lying between Pinnacle Mountain and Mount

Temple. Though this kind of formation is common in the Rockies, no other similar towers have been discovered that surpass these in height and symmetry in any accessible part of the mountains. The entire mountain is more or less eroded into sharp ridges, steep precipices, and lofty pinnacles. The largest rock tower is conspicuous from many parts of the valley, and is roughly estimated to be 300 feet high.

(*a*) Easiest route is to cross stream on south side of valley, and, ascending a thickly forested slope, skirt a pile of massive rocks (bear-den moraines), at the base of Pinnacle Mountain till the open, grassy vale is reached at base of the Sentinel Pass. From an altitude of 7,500 feet, where the vegetation ceases to grow, there is a good view of the pinnacles. Experienced mountaineers can ascend the slope to ridge, above the conspicuous rock-towers, somewhat to right. From base of this tower (8,200 feet) a very fine view is obtained, but caution is necessary, as even the largest stones are unstable and treacherous.

(*b*) An alternative route is by ascending slopes above snow patch on west base of Pinnacle Mountain. This leads to ridge (8,500 feet), whence a fine view of the pinnacles is obtained from above. On account of loose stones, experienced climbers only should make this ascent.

* *Sentinel Pass,* so called from the pinnacles, resembling sentinels, on Pinnacle Mountain, and very appropriate also from a small tower on the pass itself, which has a remarkable resemblance to a sentinel on guard, may be reached from Paradise Valley without much difficulty. From the summit (8,556 feet) there is * *an excellent view* of Mount Fay and the lower part of the Wenkchemna Range. The view into Paradise Valley is wild and awe-inspiring.

At almost equidistant points on the circumference of a circle about three miles in diameter, the center being the Horseshoe Glacier in Paradise Valley, there are five passes: the Mitre (8,500 feet), Wastach (8,338 feet), Wenkchemna (8,521 feet), Opabin (8,450 feet), and Abbot (9,588 feet), connecting five different valleys. These passes, either singly

NORTH SLOPE OF PINNACLE MOUNTAIN

or in various combinations, may be used as routes for a considerable number of very interesting excursions. Camps, stationed in one or more of the several valleys, are generally necessary if more than one of these passes is crossed in a single day.

Lake Annette (named for wife of W. G. Astley, first manager of the chalet), a small, clear lake at the base of Mount Temple, may be reached from the lower part of Paradise Valley. The branch trail to this lake is hard to locate, but may be discovered as follows: It is three-fourths of an hour's walk from junction of Paradise Valley trail with main trail, or half an hour's walk below Alpine Club camp grounds. The trail commences in an open slide, exactly opposite the north sky-line of Mount Temple cliffs, and this slide may be identified as the only one in the valley, which is covered by a uniform growth of young evergreens about 25 feet high, with virgin forest on either side. The trail commences in south corner of this slide, and, once the woods are entered, is followed with ease. The stream is crossed by a log bridge, from which the lake (250 feet above) is reached in eight minutes. A recent snow-slide has thrown down many trees near the lake. Good view of cliffs of Mount Temple.

* *Mount Temple* (11,626 feet, named for Sir Richard Temple by Dr. Dawson) may be climbed conveniently from Paradise Valley by the Sentinel Pass. The climb, which is long and arduous, presents no mountaineering difficulties. From pass a route is followed somewhat to right of the arête, which comes down to the pass. At about 11,000 feet the shoulder should be turned on the left and the south slope reached, whence the top is easily attained. If unaccompanied by guides (not advisable), the tops of couloirs and other significant points should be marked by piles of stones, as the route in descent may otherwise be entirely lost. **View from the summit* is magnificent on a clear day, including the most easterly ranges of the Rockies to the Selkirks, and from the high peaks near the Athabasca Pass southward to Mount Assiniboine and the Kananaskis Pass.

**Mount Aberdeen* (10,340 feet, named for Lord Aberdeen) may be easily ascended from upper part of Paradise Valley. The ascent is monotonous and tiring, over scree slopes and rock-slides throughout. From its central location the view is remarkably fine, especially of Mounts Victoria, Lefroy, Hungabee, and the peaks of the Wenkchemna Range. A cap of snow and glacier descends the north slope, and there is at times a pool of sapphire-blue water near the actual summit.

VIEW SOUTH FROM LOWER SLOPES OF MOUNT TEMPLE

PART III.

VALLEY OF THE TEN PEAKS, MORAINE LAKE, AND CONSOLATION VALLEY.

The Valley of the Ten Peaks, seven miles long, formerly called the Wenkchemna Valley, from the Indian word meaning "ten," is the second valley east of Lake Louise. Entering the Bow Valley at about 5,700 feet, in a flat and swampy tract, the grade is slight as far as Moraine Lake, which is the central point in location and interest, and then the valley floor rises rapidly and continuously up to its end in the high and barren Wenkchemna Pass (8,521 feet). Hemmed in by the very rugged Wenkchemna Range to the south, and Mounts Temple, Pinnacle, and Eiffel on the north, it is the wildest and, in many respects, the grandest of the five valleys. A short but beautiful branch valley enters from the south near Moraine Lake, and with its two equisite lakes, a fine glacier, and Alpine peaks, is well worthy of a visit. The upper, barren part of the Valley of the Ten Peaks was discovered first, and from its forbidding nature was called Desolation Valley until the remarkable beauty of the lower part became known, whereupon the name was changed. The peaceful charm of the south branch suggested, in contrast, the name Consolation Valley before the latter change was made.

For those who have only one day for this excursion, the following plan is recommended: From chalet to Moraine Lake (8 miles—2¾ hours) on foot or horseback; luncheon at the lake, and either a short walk down the lake shore or a visit to Consolation Valley (35 minutes going; 1 hour round trip). An early start (8 to 9 o'clock) should be made, so as to take advantage of the cool part of the day and have plenty of time to enjoy the scenery upon arrival. Luncheons and horses should be arranged for the night before. It should be remembered that it is much less tiring, both for the saddle-horses and their riders, if all level parts of the trail are availed of to make good time, as otherwise the trip will be prolonged to three hours or more.

The trail starts in near east end of chalet, and, crossing outlet stream by a corduroy bridge, ascends rapidly till (6 minutes) the trail divides, the left-hand branch being followed. (For details of trail, see page 21.) The turbulent Paradise stream is crossed by a log bridge (3 miles—1 hour). The trail now commences to ascend rather steeply, with short stretches of level grade. A small, clear stream, descending from a lake on the north side of Mount Temple, is passed in a deep ravine (1½ hours). Ascending constantly through a sparse growth of larches and pines, the trail reaches the top of a ridge (1¾ hours—6,650 feet altitude). There is now a short stretch amongst morainic ridges which form the clear-cut entrance to the valley, the trail turns to the right, begins to descend, and there is unfolded, quite suddenly—

***A magnificent view of the entire lower Valley of the Ten Peaks, with Moraine Lake* in the distance. The trail is high above the valley, in which the broad and sluggish stream is seen far below. The blue-green Moraine Lake, three miles distant, looks deceptively near, and beyond it the Wenkchemna Range and Consolation Valley are seen in a most impressive panorama. The trail, narrow and rocky, gradually slopes downward the entire distance to Moraine Lake, the interest being maintained throughout. The timber in this part of the valley was unfortunately burned in 1904, but a profusion of wild flowers, fireweed, painted cup, yarrow, and arnicas makes this place unrivaled for color in any part of the mountains from June to September. One slope is entirely covered with the cream-colored blossoms of the rather rare and interesting "erioganum." The view into Consolation Valley improves, and, at one mile from Moraine Lake, there is an excellent and striking view of the Tower of Babel and the perpendicular cliffs of Mount Babel, with its gendarmes and rock towers. The trail enters the green forest one-fourth mile from the lake, and descending, reaches the camping ground at north end of the lake.

LAKE LOUISE IN 18[illegible]

MORAINE LAKE

Alternative route is by the new carriage road, completed nearly to valley entrance in summer of 1908. Inquiry should be made at chalet as to possibility of using road as far as made, and then by trail the last three miles.

***Moraine Lake* (6,190 feet), a most beautiful sheet of water, named from the ancient moraine which forms the lake, is slightly over a mile in length, intensely blue or blue green in color, very clear, and fed by cold streams and underground springs, which may be seen bubbling up in many places under the water. The lake abounds in brook trout from 1 to 2 pounds in weight, but has been over-fished in late years. The lake is enclosed on the south side by the imposing cliffs of the Tower of Babel, Mount Babel, Mount Fay, and several others of the Wenkchemna Range. The west shore is a slope of green timber, in which may be seen still standing the tall spars of dead timber, relics of a prehistoric forest fire, which apparently visited this valley about 75 years ago. The early morning reflections, which rarely last after 9 o'clock, are marvelous. The best view of the lake, which the visitor should not fail to see, is from the top of the moraine. Cross outlet stream 100 yards below old log-jam, where there is a footbridge. Ascending trail a few yards, turn to right and climb slope where it is easily accessible. This moraine, 80 feet above the lake, gives a magnificent view, far more impressive than from below, and when the lake is calm the impression is of looking into a sapphire bowl deep-set among rugged mountains.

**Another view* of the lake, quite different in character, is obtained by walking 10 minutes along trail which skirts the north shore of the lake. The first part is somewhat swampy, but in five minutes a small stream is crossed, and from here on the trail is dry and excellent, through a beautiful grove of young trees, for one-fourth mile, till another stream is reached and a gravel flat covered with willows. This is an excellent place to take luncheon. The lake here is broader, and through a deep gorge southwards the sharp, snowy crest of Mount Fay, unseen at other end of lake, looms grandly. From the vertical precipices on either side, crowned with walls of ice, waterfalls descend in slender silvery streams, and avalanches fall, on warm days, with some frequency.

Consolation Valley, a typical Alpine valley of great beauty, may be hurriedly visited in 1 to 1½ hours from Moraine Lake. The trail crosses the outlet stream about 100 yards below the lake, and, winding through the ridges of the moraine, crosses the rock slides beyond and commences to ascend the forested lower slopes of the Tower of Babel. After traversing the green forest for a short distance the noise of rushing water is heard, and the trail, bordered by wild flowers, approaches the roaring torrent that descends from Consolation Valley. Bending to the right, the stream is soon left behind as the trail, having gained the level of the valley opening, passes through a thin growth of young trees. On the right is the Tower of Babel, and glimpses of the snowy peaks at the valley end are enjoyed through the scattered spruces. The trail enters a strip of heavy forest and reaches an open meadow. The visitor should here turn to the left, to the banks of the broad, sluggish stream, where (35 minutes) there is an—

***Excellent view of Consolation Valley,* the forests and mountains being reflected in the quiet water. The whole scene is dominated by Mount Bident (10,109 feet), a serrated, glacier-clad peak of remarkable beauty. This view combines every element of grandeur and beauty characteristic of the Canadian Rockies—vertical cliffs and glaciers, forests, meadows, and a water surface, the reflections in which give an added interest to the surroundings.

**The Consolation lakes,* about one-half mile up the valley, may be reached by the trail which follows the west side of stream. The first lake is more than half a mile long, very deep and clear, and enclosed by rock-slides and forests. In this lake, as in the stream below, there is excellent fly fishing for brook trout averaging about one pound or less. At one point on the west shore a bank of perpetual snow may be used as a landing place for the fish, making an exceptionally rare combination. Unfortunately, these lakes are now being excessively fished by irresponsible parties, whose selfish interests will result in the termination of legitimate sport.

CONSOLATION VALLEY AND MOUNT BIDENT

within a short time unless stringent game laws are enforced. Trail crosses and is now on east bank of lake.

Separated about 75 yards by a rock-slide through which the water percolates, the upper lake is considerably smaller and rests against the snout of a glacier. By ascending the valley for a short distance and climbing the moraine, this glacier may be easily reached and traversed safely for a considerable distance. The glacier is unusually interesting, as the medial moraines are clearly marked, and there are several remarkable surface lakelets and other features deserving attention.

The Tower of Babel (7,580 feet) may be ascended by way of the steep and grassy slopes near middle of the valley. In order to avoid heavy scrub and timber, the route should be studied out in detail before commencing the ascent. The top of the Tower of Babel is unusually flat, though the horizontal ledges have been split into deep and rather ominous-looking cracks by ages of exposure. **The view*, especially of Moraine Lake, 1,400 feet below, is interesting and unusual. By following the ridge southwards, climbers of a little experience may easily reach a point several hundred feet higher and get a good view of the towers and gendarmes on the east face of Mount Babel. The visitor is cautioned not to attempt the descent to Moraine Lake over the scree slopes on the west side of the Tower of Babel, as this is excessively dangerous on account of frequent rock-falls from the rapidly disintegrating gorges above.

The pass at end of Consolation Valley (8,300 feet) may be used by fair climbers as a route to the very beautiful and interesting region round and near Boom Lake, where there is excellent fishing. For a prolonged sojourn, a camping outfit should be sent round by way of the Vermilion Pass.

The Wenkchemna Lake and upper Valley of the Ten Peaks may be reached over a fair trail from Moraine Lake in 1½ to 2 hours. The trail skirts the north shore of the lake, crossing a stream from Larch Valley about half way down the lake, and then pursuing a rough but picturesque route along the forest-lined and rock-ledged banks to near end of lake (20 minutes), where the inlet stream bursts from its banks of willows and enters the lake. This stream is remarkable as being one of the few, if not the only, *clear* glacial stream in the mountains. The trail ascends through light timber and open swampy meadows in successive sharp pitches, about 2½ miles, always to the right of and above the Wenkchemna Glacier, the moraines of which may be seen occasionally through the trees. At this point (50 minutes from lake end) the forest opens out, and for nearly a mile the trail traverses a beautiful Alpine park, which, in late July and August, is a veritable garden of wild flowers unsurpassed in the mountains. Through open stretches the eye wanders over extensive fields of lilac-colored asters of vigorous growth, scarlet painted cups, white valerian, and yellow arnicas. Through the trees the cliffs and serrated peaks of the Wenkchemna Range rise superbly. Skirting the base of a snow-slide, covered with bent spruce-scrub and broken stones, the trail crosses a grove of larch trees, gradually ascending to a high gulch or pass with patches of perpetual snow (where horses can proceed no farther), then descends about 100 feet to the small but remarkable—

***Wenkchemna Lake* (about 7,500 feet), a mountain tarn entirely surrounded by broken masses of rock of great size. In this wild setting, only relieved on the north by a green slope and a few trees, the little lake nestles and peacefully reflects the stupendous mountains that surround it. The ice-cold and green water, clear as crystal, comes from melting snowbanks, which in several places border the lake shore all summer. The entire Wenkchemna Range, from Mount Fay to the Wenkchemna Pass, including the gigantic wedge-shaped Mount Deltaform, with its glacial couloir, is seen to the east and south. Apparently near, but about 3 miles distant, is the Wenkchemna Pass, leading over into Prospectors Valley. The visitor is advised to follow north shore, crossing the rocks by which a smaller lake is cut off on the right, and then cross the immense masses of rock till the entire range can be seen reflected in the water. To the west is the great rock peak of Mount Hungabee, seen through a gap, and to the right, forming part of the valley boundary, are two unnamed peaks, the most distant of which, with flat summit

WENKCHEMNA LAKE LOOKING SOUTHEAST

and deeply eroded gullies, resembles an ancient ruin. Both this and the unnamed peak to the right are marvels of light and shade in the afternoon sun. Somewhat more to the right is the Wastach Pass, an easy route into Paradise Valley, and to the north Eiffel Peak, here showing to great advantage, its vast slopes being crowned by a rounded mass of nearly vertical cliffs rent into several deep-cut needles. The lake, while not much more than one-fourth mile in diameter, is so surrounded by rough rocks that it requires nearly an hour to walk around it. The visitor can ascend the valley through open, treeless moors and over snow patches of increasing size to summit of Wenkchemna Pass (1 hour—8,521 feet). This is part of the continental watershed, the valley to the west (Prospectors Valley) lying in British Columbia, and the water draining into the Pacific by the Vermilion River. Across the valley there is a long rock wall, culminating in the sharp-peaked Mount Biddle on the right. In the valley, nearly 1,000 feet below, a remarkable limestone tower is clearly seen. This pass may be used as a route to the glacier-bearing Opabin Pass, which leads to Lake O'Hara, as well as the ascent of Mounts Hungabee and Deltaform.

The Wenkchemna Glacier, which occupies the southern part of this valley, lying against the foot of several of the Ten Peaks, has many remarkable features. Being of the piedmont type, its breadth, 3 miles, is proportionally great for its length, which is only from one-half to one mile. A large part of its surface is completely buried in morainic debris, which serves as a protecting cover to the inert and sluggish mass of ice beneath. The clear and limpid glacial stream descending into Moraine Lake proves how little eroding action is going on under this extraordinary glacier. About 2½ miles above Moraine Lake the glacier is apparently advancing, and the forest is being overridden and buried in the frontal moraines. (For fuller details, see "Glaciers of the Canadian Rockies and Selkirks," by Dr. William Hittell Sherzer.)

Larch Valley and Sentinel Pass.—Larch Valley is a short but beautiful upland vale between the south slopes of Mount Temple and a long ridge which sweeps down in a graceful curve from the base of Eiffel Peak. It heads in the Sentinel Pass and a rocky amphitheater between Mount Pinnacle and Eiffel Peak. It contains three small lakes and commands fine views of the rugged mountains to the south, and is a beautiful place in itself, though involving a hard climb to reach.

From Moraine Lake the ascent (800 feet) is most conveniently made by following the old fire-guard, which is picked up about one-half mile from the lake on trail from chalet. From the larch groves in the upper Valley of the Ten Peaks the open ridge to the north is easily ascended, the climb here being, of course, much less than from Moraine Lake.

From the nearly barren ridges above the highest lake the route by which Mount Temple is climbed is easily made out. Note a natural bridge in cliffs on side of Mount Temple, about 3,000 feet above the valley. From lake the ascent of Sentinel Pass is easily made in 30 to 40 minutes. The steep slopes should be climbed somewhat to right on Mount Temple side, avoiding the apparently easier cliffs to the left.

Eiffel Peak and the *mountain west of the Wastach Pass* may be ascended without difficulty and without guides from camps in upper part of the Valley of the Ten Peaks.

Peak northeast of Mount Temple may be easily ascended from the ridge where the trail, entering the Valley of the Ten Peaks, makes a sharp turn to the right and Moraine Lake is first seen. Near the summit the east arête is made of a remarkable and rather unusual slate formation. On and near the summit there is disclosed a very excellent—

* *View of the Wenkchemna Range and Consolation Valley* and a broad outlook over the Bow and Pipestone Valleys.

About 800 feet below, nestling in a secluded rocky vale, may be seen a blue lake about one-third of a mile long, with its end against the cliffs of Mount Temple. The descent may be varied by keeping somewhat more to the west and coming down on a small pool at base of east cliffs of Mount Temple, where a high waterfall descends upon a bank of snow.

MOUNT DELTAFORM

From high bench, a mile farther south on lower slopes of Mount Temple (reached by leaving Moraine Lake trail one mile north of lake and climbing the steep but open grassy slopes), a magnificent view of Consolation Valley, the Tower of Babel, and Mount Babel may be enjoyed. From this point the view is unsurpassed for a certain wild and imposing grandeur. Above the fire-swept zone there are scree slopes rising several hundred feet higher to the vertical cliffs of Mount Temple, where the ambitious climber may obtain more extensive and possibly more impressive views.

The circuit of Moraine Lake is a trip recommended to those camping at or near the lake. Follow trail to end of lake and cross the inlet stream about one-fourth of a mile above its mouth. Traverse woods in direction of Mount Fay to the sharp-pointed end of lake, where a small, muddy stream enters. Cross the flats and skirt shore of lake northwards, through thick scrub and timber (trying and difficult), climbing to base of cliffs about 200 feet above. By following the now easy route along the ledges, inspiring and constantly varied views of cliff and lake scenery may be enjoyed. In one place there is a small cave where the roof overhangs a considerable distance. The ledge leads out to slides, the first of which should be crossed rapidly for fear of falling stones, the others being safe.

None of *the peaks of the Wenkchemna Range* should be attempted without Swiss guides and a well-prepared camping outfit. The guides stationed at the chalet will give all necessary information to those planning to climb any of these difficult but interesting Alpine peaks. Mount Deltaform is the most dangerous of this group, and has only been climbed once. Mount Hungabee, which, with Deltaform, are attacked by way of the Wenkchemna Pass, is similar in difficulty. These peaks, ranking with the most difficult of European mountains, should be attempted only by expert climbers, with a well-organized party of Swiss guides and under favorable weather conditions.

To Moraine Lake, Wenkchemna Lake, and over the Wastach Pass, thence by Paradise Valley to chalet (about 25 miles) is a magnificent walk, with no climbing difficulties, but is only recommended to strong walkers, as being somewhat too long for the average ability. This trip might be made easier by taking saddle horses as far as Moraine Lake. It is a combination of several excursions described elsewhere, and gives a great variety of highly interesting scenery.

LAKE O'HARA AND MOUNT LEFROY

PART IV.

LAKE O'HARA REGION.

Lake O'Hara and the upper part of Cataract Brook Valley are unexcelled by any part of the mountains in variety and beauty of scenery. Though somewhat less rugged than the Valley of the Ten Peaks, and less grand than Paradise Valley or Lake Louise, this region appeals more strongly to the artist and the camper. Meadows, forests, and lakes are here combined in pleasing variety in a broad valley surrounded by mountains of strong and impressive outlines. It is possible to visit Lake O'Hara hurriedly in one day, but the visitor is strongly advised to spend at least one night in camp, so as to do at least partial justice to this very interesting region.

The lake may be visited most conveniently by taking the train to Hector station, where horses and guides, previously arranged for, should be met. Hector may also be reached by direct trail from the chalet (on foot or horseback) in three hours. This trail commences north of the chalet, and, following the original road to Laggan a short way, branches off to the left and skirts the base of Mount Saint Piran and Popes Peak, coming out east of Sink Lake, and then following the railroad track to Hector.

From *Hector* (5,190 feet) the trail ascends the east bank of Cataract Brook, a roaring torrent of clear water which enters *Wapta Lake* a few yards below. The scenery is barren and desolate, but the attention is soon attracted by the canyon-like walls of the torrential stream. In a short quarter hour, after a steady ascent of 400 feet, the top of the ridge is reached, and the trail now keeps a nearly level grade for the next three miles. The valley is entered through burnt timber and muskegs, and the two *Narao Lakes* and the sluggish stream are seen on the right (30 minutes). A grassy meadow is reached (50 minutes), whence may be seen a part of Mount Victoria, glacier-crowned, on the left, the Wiwaxy Peaks, Mount Schaffer, and Mount Odaray, far up the valley, with Cathedral Mountain on the right. Shortly after this a gravel wash is crossed, where the trail must be followed with care, and on the other side of this the burnt timber is left behind (one hour) and the green timber entered. The scenery now rapidly improves, and, in fact, is an ascending climax all the way to Lake O'Hara. The trail crosses the clear, rushing stream, which affords a succession of delightful views for several miles, by a bridge (one hour and fifty minutes; 6,150 feet), and continues on west side of stream. The ascent is rapid up to some meadows near the base of the Wiwaxy Peaks, and at length (three hours) the first glimpse of the small lake near outlet of Lake O'Hara is caught, and in five minutes more, after passing through a grove of spruces, the richly colored waters of Lake O'Hara appear directly ahead.

** *Lake O'Hara* (6,664 feet), claimed by some to be the most beautiful lake in the world of its size, is an emerald-green sheet of water nearly a mile long, almost completely enclosed by forest-clad mountains. It was named after Col. R. O'Hara, one of the early visitors to this region. The stupendous cliffs of Mount Victoria and Mount Lefroy rise above it to the north and east, and the sharp peaks of Mounts Yukness, Hungabee, and Biddle make the views to the south even more interesting, while from the farther shores the circle is completed by Mounts Schaffer, Odaray, Stephen, Cathedral, and the Wiwaxy Peaks, making the total of high and interesting peaks visible from this small lake far greater than from most others of much larger size. The first opportunity should be taken to walk along the south shore (on the right) nearly or quite to the end of the lake (30 minutes round trip). A faint game trail indicates the way, which is comparatively easy. The visitor should also cross the outlet stream (felled trees) and walk a few hundred yards along the north shore, in order to get an excellent view of the Opabin Pass, with Mounts Hungabee and Biddle on either side.

LAKE O'HARA AND OPABIN PASS

The following three excursions will give the best idea of the surrounding region:

1. *To top of cliffs below Opabin Pass,* by way of meadows and upper lakes, returning by gulch on east side to Lake O'Hara (1½ hours).

2. *To Lake Oesa* by north shore, returning by south shore (2¾ hours).

3. *To Lake McArthur* (3 miles; 2¼ hours round trip).

1. The lower part of the Opabin Pass is well worthy of a visit, and the following route is the most interesting: Take the main trail, which turns west a few yards from the lake, and which, ascending a low ridge, leads down to a meadow (4 minutes) completely surrounded by forest. This beautiful place, whence the views of the mountains are most interesting, was Colonel O'Hara's favorite camping ground. The trail turns to the right, following a mossy stream, but the visitor will see an open meadow to the southeast in the direction of the Opabin Pass, following which and then crossing a narrow bit of forest he will soon arrive on the shore of a long, narrow lake (12 minutes). This sheet of water, with its rocky forest-lined shores, while not comparable to Lake O'Hara, would in any other situation command unbounded admiration. The walking (along west shore) amongst the rocks and fallen trees is trying, but the end is reached in a short one-fourth hour. From this end of lake, where a rock-slide from Mount Schaffer is crossed, an easy route is now observed up the grassy slopes, on right of the rounded dome of rock directly ahead. The top of bench (7,240 feet) is reached after 20 minutes of climbing. An extensive high valley, with ridges and ledges of shaly rocks, is now disclosed, leading up to the snow-covered base of the *Opabin* (Indian word meaning "rocky") *Pass*. Some little time can be pleasantly spent wandering around amongst the attractions of this delightful region. The cliff edge also should be approached, whence a very **fine view of the Wiwaxy Peaks* and the lakes in the valley below may be enjoyed. The descent is made on the east side of the rounded dome, where a steep grass slope, east of a rock-slide and west of the stream, may be found. (Do not go too far to right.) This

stream divides some distance above the lake, and enters by two branches about 150 yards apart.

2. **Lake Oesa* may be visited (a very interesting excursion), by following the north shore of the lake, as a variation of route, to near its end, where the steep shores and the open slopes above, suggest ascending so as to come out above the slides of immense rocks. A small, green lakelet (6,990 feet), without apparent outlet, which, however, supplies the waterfall at end of Lake O'Hara, is reached (50 minutes). Above this, the stream becomes a boisterous torrent, hemmed in by rock-slides and wild gorges. The stream is crossed and followed closely on west bank, the scenery being most interesting and constantly varied. Leaving the stream for a short time, a scramble is made up an interesting rock-slide to right of cliffs, which appear to block the way. Above these (7,350 feet), an interesting vale of ledges is entered and easily ascended, passing a green pool on the left (1¼ hours). The vegetation now almost totally disappears, the valley assumes a wild and desolate aspect, and in a short time the barren shores of *Oesa* (Indian word meaning "ice") *Lake* is reached (7,398 feet—1½ hours). This region is strongly marked by the striations and grooves of former glacial action. From this point the route to *Abbot Pass* may be clearly seen leading up to a deep cleft between Mounts Lefroy and Victoria. The return should be varied by descending the slopes *west* of the waterfall at Lake O'Hara and following the west shore of same.

3. **Lake McArthur* is a remarkable lake, easily reached from Lake O'Hara in one hour. The trail near lake which leads to Colonel O'Hara's meadow should be followed. This is an old Indian trail, a relic of prehistoric days, when the Kootenay Indians used it as a route to reach and trade with the various tribes on the plains. It is hard to follow through the meadow, but with care it will be seen to lead west to a shallow gulch, and is well marked in the forest. Ascending steep pitches, the top of first bench is reached (20 minutes), and the first view of the pyramidal Mount Owen is seen to the west. The trail, becoming faint as it crosses open meadows, passes a small lake on the left. Cross these

LAKE McARTHUR AND MOUNT BIDDLE

meadows towards a rocky gulch somewhat to the right, where there is a small stream. The trail is on the south or left of this. The summit of pass (7,110 feet) is reached (35 minutes) and the valley opening, where Lake McArthur is situated, may be seen to the south between Mount Schaffer and Park Mountain, about three-fourths mile ahead. (Just before reaching summit of pass there is a small meadow filled with great rocks. One particularly large boulder, with strata tilted up to the west and a loose slab on east side, should be passed on east side, turning to left.) The slopes are descended, attention being given to the route taken, but keeping in general somewhat to the left till the base of gulch is reached, where, ascending by a well-marked goat trail to top of a rocky bench, there is no further difficulty. The visitor should now mark his route by piles of stones on conspicuous points, so as to return by same gulch. Crossing the barren, rolling ridges of this high valley, the sapphire-blue waters of Lake McArthur are soon seen in the distance.

**Lake McArthur* (7,359 feet; named for J. J. McArthur, who discovered it), one of the largest lakes at such high altitude in the mountains, lies in a short Alpine valley between Mount Schaffer and Park Mountain, and is overhung on the southeast by Mount Biddle (10,876 feet). A large glacier descends from the latter peak and enters the lake. This lake is remarkable for the extreme coldness and clearness of the water and for its wonderful sapphire-blue color, having no trace of the blue-green characteristic of most Rocky Mountain lakes. Owing to its high altitude, it is covered with ice most of the year, and even in midsummer floating masses of ice drift over its surface. On the extreme right-hand corner there are some whirlpools where the lake seeks a subterranean outlet. Towards the left there are some interesting cliffs and rocks, whence the best view of the lake is obtained. Visitors should carefully examine the surrounding mountain slopes, especially the western slopes of Mount Schaffer, as this is a favorite haunt of the Rocky Mountain goat, and it is more than likely that some of these interesting animals may be seen.

LIST OF INTERESTING BOOKS AND PUBLICATIONS.

"Among the Selkirk Glaciers." (In Chapter XVII is an interesting account of an early visit to Lake Louise.) W. S. Green. Macmillan & Co., London.

"Climbs and Exploration in the Canadian Rockies." H. E. M. Stutfield and J. Norman Collie. Longmans, Green & Co., London.

"The Canadian Alpine Journal." Published annually by the Alpine Club of Canada.

"In the Heart of the Canadian Rockies." James Outram. The Macmillan Co., New York.

"The Rockies of Canada." Walter Dwight Wilcox. G. P. Putnam's Sons, New York.

SPECIAL SUBJECTS.

"Alpine Flora of the Canadian Rocky Mountains." Stewardson Brown and Mrs. Charles Schaffer. G. P. Putnam's Sons, New York.

"Mountain Wild Flowers of Canada." Julia W. Henshaw. William Briggs, Toronto.

"Glaciers of the Canadian Rockies and Selkirks." William Hittell Sherzer. Published by the Smithsonian Institution.

Attention is also called to the very interesting "Preliminary Report," by Dr. Dawson, and the exceedingly rare "Palliser Papers." A "Report on the Geological Structure of the Rocky Mountains," by R. G. McConnell, is of interest to geologists. Besides Appleton's Canadian Guide Book and Bædeker's Guide to Canada, there have been published in the last few years a great number of articles relating to this and other parts of the Canadian Rockies in the following magazines: Appalachia, the Alpine Journal, the Geographical Journal, the National Geographic Magazine, the Journal of Geology, etc.

LIST OF BOOKS RELATING TO THE HISTORY OR EARLY EXPLORATION OF THE CANADIAN ROCKIES.

"The Northwest Passage by Land." Viscount Milton and Dr. Cheadle.

"Canada on the Pacific." Charles Horetzky.

"Voyages." Sir Alexander Mackenzie.

"Overland Journey Round the World." Sir George Simpson.

"Saskatchewan and the Rocky Mountains." Earl of Southesk.

"Ocean to Ocean." Rev. George M. Grant.

"Adventures on the Columbia River." Ross Cox.

Canadian Pacific Railway Reports.

NOTES FOR WALKERS AND CLIMBERS.

The exhilaration of the mountain air and the entire change of climate have a tendency to make visitors overestimate their physical powers, especially in the first excursions taken. Travelers of a sedentary life should be extremely careful not to attempt too much, and even those who have hardened their muscles by golf, tennis, and other outdoor sports must gradually accustom the heart and lungs to a rarefied air and an entirely new kind of exercise. By adopting the following ideas the best results will ensue:

1st. The first few days should be devoted to walks along roads or trails that are comparatively level. If any small ascents are made they should be done leisurely. It is highly important to have a pair of comfortable, thick-soled shoes, well filled with hob-nails or Swiss Alpine nails. A pair of canvas or other leggins and an old suit are advisable.

2nd. The secret of mountain climbing, thoroughly understood by the Swiss guides, is "festina lente"—making haste slowly. This is especially true during the first hour of a day's climb, when the heart is becoming accustomed to the extra exertion in a rarefied atmosphere. More physical energy may be destroyed by rushing along during the first half hour than in any two or three subsequent hours of the day's work. Remember that the chief object of any excursion is to enjoy the grandeur and beauty unfolded along the way. Nothing so plainly marks the tyro as record-breaking proclivities.

3rd. In all big climbs the exertion expended in descent is more tiring than in the ascent. The mountaineering difficulties also are more serious, as the footholds are less easily seen. Have the moral courage and good judgment to turn back in time to reach camp or hotel before nightfall.

AN APPEAL FOR THE PRESERVATION OF NATURAL BEAUTY.

There is a strong movement on foot both in the United States and in Canada towards the preservation of natural scenery. Those who have had the rare pleasure of being a first visitor to any of these beautiful lakes of the Canadian Rockies and seen nature in her primitive glory may thoroughly appreciate the destructive effect of the ordinary camp. Trees are ruthlessly cut down or destroyed by camp fires, branches are hacked off for fire-wood or bedding, the bark is blazed from trees and replaced by a multitude of names of those seeking a misguided notoriety, while the green carpet of grass and beautiful Alpine flowers is changed into a waste of empty cans and broken glass. Unfortunately such rude transformations almost invariably occur at the most attractive point of view, making a strong contrast between nature's sublimity and man's heedlessness.

It would be a most excellent thing if, through the efforts of the officials of the Rocky Mountains Park of Canada, supplemented by an aroused public sentiment, working through clubs organized for the purpose, a rational system of camp supervision were decidod upon and rules enforced by mounted forest rangers. Every forest-lined lake or stream; every lonely mountain tarn or Alpine meadow—in fact, all places in any part of the world where beauty and grandeur is preëminent over utility—are a portion of the natural heritage and inalienable rights of all the people. Any one who destroys this beauty and takes away some of the pleasure that belongs to those who come after is, in a sense, a robber and a malefactor, no less than if he should cut down the shade trees and ruin the gardens of a private home. It must be remembered, however, that the effectiveness of this effort to preserve natural beauty depends, in its final working out, upon the individual visitor, who, unfortunately, is often overwhelmed with the limitlessness of the forests and the grandeur of nature and is led, moreover, to admire and overestimate the apparent knowledge and experience of his guides, little realizing that in many cases these selfsame guides may be spending their first season in the mountains, having little or no appreciation of that beauty which they are so ready to destroy.

A FEW CAMP "DON'TS."

Don't place your camp on the borders of lakes or streams or in any fine scenic viewpoint.

Don't cut down and mutilate trees. Don't write your name at the expense of the bark and often the life of a noble tree.

Don't build a camp-fire in mossy ground or heavy brush.

When breaking camp, don't fail to pour on water until the last spark, deep down in the ground, is extinguished.

Don't let your camp resemble a rubbish pile. Burn up papers and refuse and hide all empty cans in a deep part of the forest, leaving your camp-ground as near like what you found it as possible.

Don't be a record-breaker in killing game or fish. Sentiment has changed in recent years, and many an alleged sportsman degenerates into a "game-hog" when he takes more life than is reasonable or necessary.

A little knowledge of weather conditions will often save valuable days or prevent disastrous results in mountain-climbing. Those who understand the weather in any part of our continent will soon notice the similarity of changes in the Canadian Rockies, affected only in a slight degree by the mountains themselves.

The cyclonic whirls, from 500 to 1,000 miles in diameter, move in from the Pacific at fairly regular intervals throughout the year. Upon their intensity and rate of motion depend the amount of rain or snow that falls and the duration of each storm. The approach of one of these storms is almost invariably indicated at least 24 hours in advance by the feathery cirrus clouds, which first appear on the western horizon and cross the sky rapidly, gradually becoming thicker and forming a hazy veil over part, or all, of the sky. The wind shifts to south or southwest, and cumulus clouds begin to form, moving in the same direction. If the wind is very strong and hot, the storm usually breaks in a series of local thunder-storms, sometimes followed by general rains. If the wind is less strong, or from the east, and the temperature is low, the storm begins in a gentle, steady shower and lasts from two to three days or more. The passage of the center of the storm is marked by a rising barometer, the clouds lifting somewhat, the temperature dropping, while new snow falls upon the mountains or even in the valleys. It often rains or snows, however, many hours after the barometer begins to rise. In general, the more rapid the approach of a storm and the heavier the rainfall, the more rapid is the clearing. West or northwesterly winds are an almost certain indication of clearing. Cold, frosty nights are a sign of continued fair weather. In mid-summer, even during the finest weather, the valleys are often filled with heavy fog in the early morning, and many excursions are missed from apprehension of an approaching storm. Such fogs may be distinguished by the fact that they entirely cover the valley to its lowest parts, while the air is cold and there is no actual fall of mist or rain. At about 9 or 10 o'clock these fogs begin to rise and melt away in the sun, showing a perfectly clear sky above.

The following average weekly thermometer records were taken by the author at Lake Louise in the summer of 1894:

	7–9 A. M.	1–2 P.M.	7–9 P. M.
July 25–31, inclusive	49	59	51
August 1–7, inclusive..................	53	69	56
August 8–15, inclusive..................	49	60	50
August 16–22, inclusive.................	56	74	57
August 23–29, inclusive.................	50	65	51
August 30-September 5, inclusive........	42	53	43
September 6–12, inclusive...............	41	55	44
September 13–19, inclusive..............	37	46	39
September 20–26, inclusive..............	31	41	31
September 27–October 3, inclusive.......	33	42	32
Hottest day, August 19..................	57	78	56
Coldest day, August 28..................	35	41	40

NOTES ON PROVISIONS, ETC., FOR A CAMPING TRIP.

One of the most difficult things about outfitting a camping trip, whether it lasts a few days or several months, is to know the quantity and quality of the provisions. On a short trip this is relatively unimportant, but for anything over a week or ten days a careful study should be made of the "grub-pile." In general, the camper-out in the Canadian Rockies is exposed to constant low temperatures, except at midday, and, moreover, he is naturally taking an abnormal amount of exercise. For these reasons the heat-producing foods—fat pork, bacon, chocolate, and sweets—and the energy-producing foods, such as beans and bacon, should form a large part of the commissariat. Fresh meats and vegetables being out of the question, a supply of dried fruits (apricots, apples, prunes, and peaches) and desiccated vegetables is absolutely necessary. From my own experience the following table is approximately correct. Weights are given in pounds and decimals. In order to find required quantity, multiply together the number of individuals in the party, the number of days, and the given decimals:

Absolute Necessaries on Long Trips.		*Advisable.*	
Bacon	1.00	Cocoa	.03
Flour	1.00	Corn meal	.10
Baking powder	.05	Rice	.05
Salt	.05	Hardtack	.10
Sugar	.20	Onions	.20
Tea	.04	Preserves and jam.	
Coffee	.08	Pickles.	
Beans	.20	Sweet chocolate.	
Dried fruit	.07	Self-rising flour and syrup.	

On short trips many of these items may be left out and their places taken by fresh bread, canned meats, sardines, eggs, butter, and potatoes.

One of the most important items in camp cooking is fresh bread made with baking powder and locally called "bannock." Only about one in ten of all the camp cooks has the faintest conception of making a light and appetizing bannock. The following directions will give excellent results: Measure out twenty parts (twenty tablespoonfuls, for instance) of flour, one of baking powder, and one-half of one of salt, and mix thoroughly and intimately in a bowl. Make a depression in the flour and add cold water, stirring the flour in from the sides till nearly all the flour is mixed and the dough, though light and spongy, is moderately firm. Stir or knead as little as possible. Take two frying pans, well greased with bacon fat or else well sprinkled with dry flour, and make a thin loaf covering entire bottom of pan, from one-half to three-quarters of an inch thick. Heat over a bright, clear camp fire, so as to cook a bottom crust without burning; then stand the skillet up on chips to keep off the ashes, by means of willow branches run through hole in handle, so that the upper surface is exposed to intense radiation from a clear fire. The loaves should rise and brown rapidly and be turned frequently, so as to be thoroughly cooked in 20 minutes. By improvising an oven from an old cracker box, and placing hot cinders below and above, the lightest soda biscuits may be cooked in a few moments. By adding lard or bacon fat, bannocks can be made lighter. By adding more water the latter can be used for making pancakes, which should be cooked in very clean pans over a *small,* hot fire. The batter should not stand, but be mixed as used. Self-rising buckwheat flour mixed with a little water can be rapidly converted into pancakes which are delicious with maple syrup.

Camp fires for cooking should be as small as possible, and only very dry wood used, as the heat and smoke are trying at best.

Owing to the altitude, everything cooked by boiling requires two or three times as long to cook as in other parts of the country. This is especially true of onions, beans, etc. Coffee, and even tea, may be boiled a short time, as boiling only corresponds to a gentle simmering at sea-level. Coffee should be carried in sealed cans, and used in a proportion of not less than one cup of ground coffee to five of water.

Nothing is more delicious than fresh trout, especially if well cooked. Thoroughly clean the fish, sprinkle salt and pepper, and roll in corn meal. Fill the frying-pan with fish, placed on their backs if freshly caught—otherwise they will curl out of the heat—and use a cupful of bacon fat or lard to cook with. Place a tin plate over the fish and cook rapidly, changing the level of frying-pan so that the hot grease reaches every part uniformly. If possible, cook one side through before turning the fish, so as not to break them up. In twenty minutes they should be thoroughly cooked, the outside showing a rich golden brown color.

A single grouse may be made to do for several men by making a "slumgullion," by cutting in small pieces and stewing in water to which is added flour, onions or potatoes, and seasoning.

The chief requisites in good camp fare is a clear, brisk fire and a little care and attention to details. It should be borne in mind that camp cooking, however it may appear at first, is really the cleanest in the world, the only kind we could watch and afterwards enjoy. The only extraneous matter that falls into camp dishes is ash and cinders from a clean, wood fire.

Experience in folding blankets will save many a sleepless night. Lay a canvas cover or some horse blankets over the spruce or balsam bed. Blankets should be laid so that the lower part may be folded under at least 25 inches. The sides are then folded under in like manner or pinned with safety-pins so as to make a kind of sleeping-bag. Other blankets are folded under in like manner in successive layers, so that there is no opening where the cold air may enter. If sleep is not otherwise possible, draw top of blanket over the head, leaving only the nose outside. This is equal to an extra blanket in warmth-giving value.

INDEX.

Spotted something that isn't quite right?

Let us know and we'll fix it for future reprints. And as a thank you for your diligence, we'll send you a free Sothis book of your choice.

publisher@sothis.press

This page intentionally left blank

Made in the USA
Monee, IL
08 November 2026

68106069R00059